Newbridge D

Animal Watch

Susan Ring

Newbridge
A Haights Cross Communications Company

Animal Watch
ISBN: 1-58273-700-2

Program Author: Dr. Brenda Parkes, Literacy Expert
Content Reviewers: Steve Shupe, Missouri Department of Conservation, Clinton, MO;
Mindy LaBranche, Cornell Laboratory of Ornithology, Ithaca, NY
Teacher Reviewer: Kerri Blaus, Sleepy Hollow Elementary School, Sleepy Hollow, IL

Written by Susan Ring

Newbridge Educational Publishing
11 East 26th Street, New York, NY 10010
www.newbridgeonline.com

Cover Photograph: Bengal tiger
Table of Contents Photograph: Scrub jay and bird-watcher

Photo Credits
Cover: Daryl Benson/Masterfile; Table of Contents page: Joe McDonald/Bruce Coleman, Inc.; page 4: Wolfgang Bayer/Bruce Coleman, Inc.; page 5: (top) Pat & Rae Hagan/Bruce Coleman, Inc., (bottom) D. Demello/Wildlife Conservation Society headquatered at the Bronx Zoo, NY; page 6: Tom Brakefield/Bruce Coleman, Inc.; page 7: John Giustina/Bruce Coleman, Inc.; page 8: AFP/CORBIS; page 9: Mark Stouffer/Animals Animals; page 11: Ken G. Preston-Mafham/Animals Animals; page 12: Michael & Patricia Fogden/Animals Animals; page 13: Jim Cronk; page 14: Phil Schermeister/CORBIS; page 15: Scott Camazine/Photo Researchers; page 16: CORBIS; page 17: Courtesy of Project PigeonWatch; page 18: (top) Frederick Atwood, (bottom) Frederick Atwood; page 20: Bill Romerhaus/IndexStock; page 21: Michael & Patricia Fogden/CORBIS; page 22: (top) Gallo Images/CORBIS, (center) Jonathan Blair/CORBIS, (bottom) Kent Knudson/CORBIS; page 23: (top) Erwin & Peggy Bauer/Bruce Coleman, Inc., (bottom) Alfred B. Thomas/Animals Animals

Map on page 10 courtesy of the Save The Tiger Fund, Washington, DC; adapted by Steve Stankiewicz

10 9 8 7 6

Table of Contents

Introduction

What do tigers eat? Why are birds different colors? How many insects are *really* out there?

Some people spend their whole lives finding answers to questions like these. They listen to animals and watch them as they eat, sleep, fight, and play. They dive under the ocean, climb up to treetops, and trek through the jungle in their search for information about animals.

This giraffe wears a special collar that will help scientists track the animal.

An underwater camera helps this scientist film a friendly manatee.

How do you think these scientists are helping the gorilla babies?

Tracking Tigers

Zoologists study animals. They ask many questions: "How do polar bears stay warm?" "Why do wolves hunt in groups?" "How do gophers live underground?" Then they get to work to find the answers.

One question is about tigers. In 1900, there were about 100,000 tigers in the world. Today there are fewer than 8,000! What happened?

Hunters kill tigers. Even more serious, the forests in which tigers live have been cut down to make room for farms and cities. How can scientists help save the **endangered** tigers?

Tigers hide among trees and tall grasses as they hunt for prey to eat. Why do you think it is important that their forest home be preserved?

Bengal tiger

First, scientists have to find out what these animals need to survive. They ask, "How many tigers live in a certain area? Is there enough space for them?" It's also important to know what tigers eat and whether there is enough food where they live.

This information can be used to determine whether the tiger's forest **habitat** is large enough and how much wildland should be preserved.

A good way to get this information is by photographing the tigers. Camera traps are set up along wildlife trails. Each camera uses a light called an infrared beam. When a tiger walks by, the camera snaps a picture.

Zoologists compare the photos from one month or year to the next. Because each tiger has its own unique stripe pattern, zoologists learn to recognize and keep track of the tigers. They can count how many different tigers live in a specific area.

Another way to gather information is by putting collars on the tigers that

This camera fits into a weatherproof box that's attached to a tree. The camera's infrared beam is invisible. It works like the "electric eye" that activates an automatic door at the supermarket.

Scientists inject tigers with medication to keep them calm while they put on the tracking collars.

send out radio signals. Researchers ride through the forest on elephants, using radio antennae to pick up the signals. They learn where the tigers are. Antennae are also set up on hills and towers.

From the radio signals, zoologists record the tigers' locations on a map. Now they can see how far tigers travel to find food and water. Some tigers cover up to 30 square miles a day! When they have to travel too far, it means that food is scarce and their survival is threatened.

TIGERS THEN AND NOW

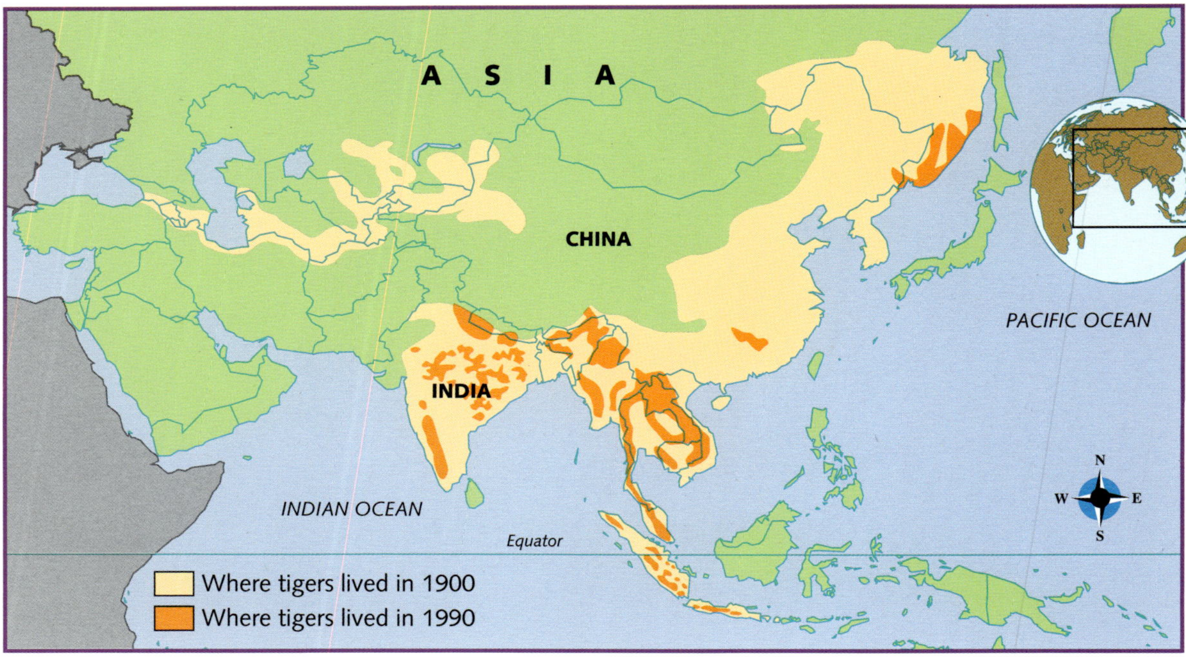

All wild tigers live in Asia. This map shows how their population has changed. What do you think this map might look like in the future? What things might people do to help save the tigers?

Camera traps and radio collars are now being used in North America to study another endangered animal, the jaguar. Like tigers, jaguars live alone and need a lot of territory.

When zoologists tackle a problem like endangerment, their work helps make other people aware of the problem. Then people who care about animals can work together to try to save them.

Discovering New Insects

There are more insects on earth than all other creatures combined. That's a lot for an **entomologist** to choose from!

Entomologists study insects. They ask all kinds of questions: "How do ants communicate with one another?" "What is the life cycle of a dragonfly?" "What makes a glowworm glow?"

Entomologists don't wait for bugs to come to them. They go find bugs wherever they are!

Different insects are attracted to different odors. Entomologists may try to attract insects with rotten fish juice, sour bananas, or other fruit, like this orange.

This leaf katydid spreads its colorful wings as a warning to predators.

Most of the world's insects live in tropical rain forests. One small area in the rain forest can have as many as 41,000 different kinds. And scientists are sure there are many more kinds of insects that haven't even been discovered yet!

The rain forest is a great place to find insects. But think about what a rain forest is like. What are some of the challenges scientists may have to overcome in order to study insects in this environment?

In the rain forest, many insects live in the treetops, or sunny forest **canopy**. They hardly ever come down. So how do scientists study them? They go all the way up to the canopy!

One method is to climb a rope to the top, then stand on a platform to view the area. In the canopy, scientists can also study the flowers or leaves where an insect lives or finds food.

Like rock climbers, these researchers wear special harnesses that enable them to climb safely up the rope. Additional safety lines connect the harnesses to the tree's branches.

Why do you think the canopy provides a good food supply for insects?

The drawers in this library contain insect specimens. What might be a good way to organize so many insects?

Entomologists bring their findings back to a place that is like an insect library. But instead of being home to books, this research center houses more bugs than you can imagine. Here scientists compare insects they've recently found with others that are already known. That's how they know if they've discovered a brand-new species.

Here are some of the questions entomologists ask to help them identify an insect:

Does the insect have wings? If so, how many?

Is the insect's waist wide or narrow?

Are the insect's antennae long or short?

A scientist who discovers a new insect gets to name it. Scientists use the insect's color, the place where it was found, or even their own name, to help them pick a name.

What might you name an insect that you've discovered?

Flamboyant flower beetle

▲ What are some questions you might ask about this insect?

Beyond Bird-Watching

Why do woodpeckers drill holes into trees? Do eagles have predators to fear? Why do some flocks of birds fly in a V-formation?

Ornithologists study birds in the wild and in zoos to find out more about them. They photograph them and record their songs. They count birds and track their populations.

Bald eagle

At this Project PigeonWatch site, the pigeons are fed at the same time each day. How might this help the pigeon-watchers?

But you don't have to be an ornithologist to find answers to questions about birds. Some answers come from people just like you!

Project PigeonWatch, organized by the Cornell Laboratory of Ornithology in Ithaca, New York, uses everyday people to learn more about pigeons.

Pigeons have lived on the earth for millions of years. All the pigeons you see outdoors are called **feral** pigeons. Feral animals, such as packs of wild dogs, usually appear in only one or two colors. This led one ornithologist to ask, "If that's true, why do feral pigeons come in so many colors?"

This is a red-bar pigeon. Its head, neck, and chest are reddish brown, and it has bars of red on its wings.

This pigeon is called a spread because its dark color is spread all over its body.

Project PigeonWatch Tally Sheet

Date 4/8/01 Time 10 a.m.

Group Name Pigeon Lookout

Recorder's Name Sam

Nearest Address to Your PigeonWatch Site Main Street and Oak Avenue Zip code 02345

PigeonWatch Site Name Steps of library

Site Number 3

Circle one in each column

How long have people been feeding pigeons at this site?

- People have been feeding pigeons here for a long time. (circled)
- We are the first to feed pigeons here.
- I don't know if pigeons have been fed here before.

Our PigeonWatch site is a:

- Street (circled)
- Yard
- Park
- Farm
- City
- Suburb (circled)
- Rural Area

Flock Count
Count the size of the flock and record it here.

~~||||~~ ~~||||~~ |||

Color Count
Next, tally the numbers of each color.

Blue-bar	Red-bar	Spread	Red	Checker	Pied	White
||||		|||	|	~~||||~~		

To answer this question, Project PigeonWatch gathers information on the colors of pigeons from people in the United States and around the world.

Teachers, kids, and parents watch pigeons and write down what they see on a tally sheet like the one above. They learn about pigeons' colors such as blue-bar and **pied**. They ask questions such as, "Are pigeons different colors and patterns depending on where they live?"

But this is just the beginning.

▲ This white pigeon is also called a dove of peace.

Recording and sharing information is an important part of finding answers. Project PigeonWatch shares everyone's data with scientists around the world. That way, all the results can be compared. The project hasn't yet come up with an explanation for the variety of colors, but the research goes on.

After watching the birds, people have asked new questions: "Why don't I see baby pigeons?" "What's the difference between a pigeon and a dove?"

What questions would you ask?

You Are a Scientist

You can be a scientist, too. Look at the animal below. How would you describe it? Compare it with the other animals you have been reading about. How is it like them? How is it different?

Collared tamandua

Other People Who

Herpetologists study reptiles such as snakes, alligators, and lizards. They also are interested in frogs and toads, which are amphibians.

Paleontologists dig for fossils, such as those of dinosaurs and plants. They also study the rocks in which fossils are found. Many paleontologists uncover information about prehistoric animals.

Veterinarians care for dogs, cats, and other animals. They give animals checkups to keep them healthy and prescribe medicine when they are sick. Some veterinarians take care of farm animals or exotic animals such as iguanas, ferrets, and parrots.

Work with Animals

Wildlife photographers take pictures of animals in their habitat, such as the pictures in this book. Many photographers wait and watch for days, and even months, to get one good photo.

Zoo nutritionists study what an animal eats in the wild so they can provide food and vitamins to match the animal's natural diet.

Websites

To find out more about tigers, insects, and birds, go to:
www.5tigers.org
www.uky.edu/Agriculture/Entomology
www.birds.cornell.edu/bow

To find out more about Project PigeonWatch, go to:
www.birds.cornell.edu/ppw

Glossary

canopy: the top level of a rain forest or any forest

endangered: threatened; at risk of becoming extinct

entomologist: a scientist who studies insects

feral: living in the wild; a word describing once-domesticated animals (and their descendants) that now live in the wild

habitat: the place where an animal naturally lives and grows

ornithologist: a scientist who studies birds

pied: having some white feathers in spots, in patches, or on flight feathers

zoologist: a scientist who studies animals

Index